My STEM Workbook 3

Understanding
Science, Technology, Engineering and Mathematics through design-process activities

Zero Hunger
Good Health and Well-being
Clean Water and Sanitation
Affordable and Clean Energy
Sustainable Cities and Communities
Climate Action
Life below Water
Life on Land

This workbook belongs to

Vinesh Chandra & Basil Slynko

Years 5–6

My STEM Workbook 3
Understanding Science, Technology, Engineering and Mathematics through design-process activities
Vinesh Chandra and Basil Slynko

Editor/Proofreader: Sandra Balonyi
Text designer: Michael Haddad
Cover designer: Michael Haddad   Concept: Basil Slynko
Illustrator: Michael Haddad

First published in Australia in 2024
Copyright ©2024 Vinesh Chandra and Basil Slynko
B. Slynko (Challenges 1, 5, 6, 8); V. Chandra (Challenges 2, 3, 4, 7)

Copyright Notice
This Work is copyright. No part of this Work may be reproduced, stored in a retrieval system, or transmitted in any form or by any means without prior written permission of the publisher. Except as permitted under the *Copyright Act 1968*, for example any fair dealing for the purposes of private study, research, criticism or review, subject to certain limitations. These limitations include: restricting the copying to a maximum of one chapter or 10% of this book, whichever is greater; providing an appropriate notice and warning with the copies of the Work disseminated; taking all reasonable steps to limit access to these copies to people authorised to receive these copies; ensuring you hold the appropriate Licences issued by the Copyright Agency Limited ("CAL"), supply a remuneration notice to CAL and pay any required fees.

For details of CAL licences and remuneration notices please contact CAL at Level 12, 66 Goulburn Street, Sydney NSW 2000.
Tel: +612 9394 7600, Fax: +612 9394 7601
Email: info@copyright.com.au
Website: www.copyright.com.au

For permission to use material from this text, please contact the publisher.
Email: **info@professorbaz.com.au**

National Library of Australia Cataloguing-in-Publication Data

Chandra, Vinesh and Slynko, Basil.
My STEM Workbook 3
Understanding Science, Technology, Engineering and Mathematics through design-process activities

ISBN: 978-0-6484052-2-1
For primary school age.

Printed in Australia
1 2 3 4 5 6 7 8 9 30 29 28 27 26 25 24

Contents

Acknowledgements iv
About the authors iv
Introduction v
Notes for the teacher vi
Connections to the Australian Curriculum viii
Notes for the student x

Challenge Activities

1 Zero Hunger 1
   – *A permaculture garden*

2 Good Health and Well-being 13
   – *A SunSmart hat*

3 Clean Water and Sanitation 25
   – *A water-filtration device*

4 Affordable and Clean Energy 37
   – *A solar solution for lighting spaces*

5 Sustainable Cities and Communities 49
   – *A kit shelter for refugees*

6 Climate Action 63
   – *A toy that moves made from waste*

7 Life below Water 75
   – *A video game on plastic-waste removal*

8 Life on Land 87
   – *A habitat for giraffes*

Some properties of materials 103
Some samples of natural and man-made materials 104
Notes and grids 105
Glossary .... Inside back cover

Acknowledgements

The authors and publisher would like to credit or acknowledge the following sources for permission to use copyright material:

Adobe Stock: p9 (hoe, hand fork, rake, shovel, pruning saw, secateurs); p31 (PPE); p52 (Cox's Bazar Refugee Camp); p58 (Phillips head screwdriver, spanner, hammer); p65 (spinning top, toy train, rocking horse); p69 (industrial snips, warrington hammer, general-purpose saw), Phillips head screwdriver; p92 (Sydney Opera House); p98 (craft knife, cordless glue gun, super glue, general-purpose saw, industrial snips).

Shutterstock: p58 (spring clamp); p64 (plastic waste).

Scratch: p79 (home page) CC-BY-SA 2.0.

Illustrations p104 by Paul Lennon.

Every attempt has been made to trace and acknowledge copyright holders. Where the attempt has been unsuccessful, the publisher welcomes information that would redress the situation.

A special "thank you!" to Sandra Balonyi and Michael Haddad for working with us on this project.

Responsibility for errors remains with the authors.

About the authors

Associate Professor Vinesh Chandra is a university lecturer and teacher with more than 40 years of experience. He has taught in Australia and several countries overseas. His teaching areas include STEM, science, mathematics and technologies. His co-authored book titled *STEM Education in the Primary School: A Teacher's Toolkit* received two awards (Educational Publishing Australia Awards, 2021). Associate Professor Chandra's team won Gold in the prestigious QS Reimagine Education Wharton Awards in 2022 for their project in STEM Education in Papua New Guinea.

Basil Slynko, aka Professor Baz, B Ed St., MA; Design and Technologies educator – primary, secondary and tertiary – in Australia and overseas; Project-based-learning advocate; Curriculum consultant; Industry Experience – Construction and Manufacturing; Author and co-author of 25 titles, including *Nelson Introducing Technology Fourth Edition*, *Nelson Technology Activity Manual Third Edition*, and *Design and Technology in Today's World: A First Look*.

What is the United Nations? Many years ago, nearly all countries in the world joined to form an organisation called the United Nations.

Introduction

Humans have always relied on Science, Technology, Engineering and Mathematics, or STEM, to find solutions to challenges. Future generations will need to holistically draw upon this within and beyond their contexts. For example, STEM knowledge and skills are vital when addressing the United Nations Sustainable Development Goals.

What are the United Nations Sustainable Development Goals?

The United Nations aims to find ways to make our planet a better place for all.

The United Nations Sustainable Development Goals are about improving the lives of people and all other living things to make our planet a better place. They are also about what we can do to care for our Earth so that future generations can also have a happy and healthy life.

My STEM Workbook 3 is part of a trilogy of STEM workbooks for primary students Years 1–6. Each workbook has eight challenges. Each addresses one of the United Nations Sustainable Development Goals (SDGs). Students apply their knowledge and skills of Science, Technologies, Engineering and Mathematics to propose solutions to contextually appropriate real-world challenges. These activities align with a range of content descriptors mandated in the Australian Curriculum Science, Technologies, and Mathematics (Version 9). However, these activities can also be implemented in other contexts, guided by other curriculum documents.

*Member states and territories of the United Nations

*Source: https://commons.wikimedia.org/wiki/File:United_Nations_(Member_States_and_Territories).svg
This work is licensed under the Creative Commons Attribution-ShareAlike 4.0 International License. To view a copy of this licence, visit http://creativecommons.org/licenses/by-sa/4.0/ or send a letter to Creative Commons, PO Box 1866, Mountain View, CA 94042, USA.

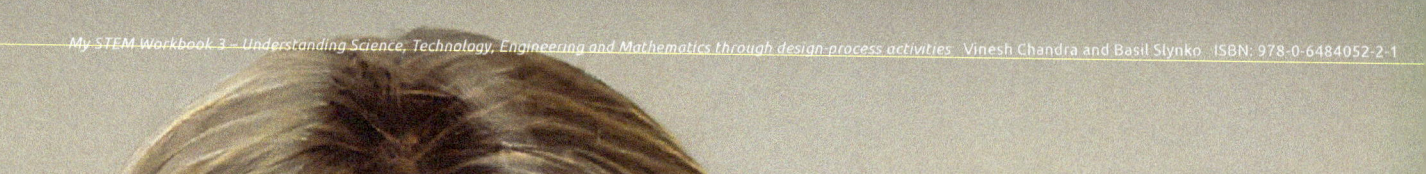

Notes for the teacher

The essence of integrated STEM education is project-based learning (PBL). It is a fun way for students to learn and teach. The real-world challenge in each activity is highly likely to interest, engage and enthuse students. *My STEM Workbook 3* comprises eight hands-on design activities, where students apply their STEM knowledge and skills to propose solutions to real-world challenges informed by the United Nations Sustainable Development Goals (SDGs). Each challenge is associated with an occupation. This should set the scene for inviting guest presenters, watching online videos and promoting class discussions.

The STEM PBL framework[1] was used to design the activities. Students need to tackle each challenge through the following steps:

> Ask → Imagine → Plan → Create → Improve

Through these steps students apply their design-thinking skills to propose solutions to real-world challenges. Workplace Health and Safety is an integral part of each activity. Students are expected to handle tools, equipment and materials with care. Teachers are also expected to reinforce the use of safety gear – that is, Personal Protective Equipment (PPE) – as needed.

The trilogy of STEM Workbooks has a website (https://mystemworkbook.com/) which presents ideas on how knowledge from the digital technologies curriculum can be embedded within each challenge. The website also has support materials, including a video commentary on each activity.

Throughout this workbook there are opportunities for students to develop their digital literacy and their knowledge and understanding of the digital technologies subject. For example, Challenge 7 specifically targets the content descriptions from the digital technologies subject. All other activities can be extended to enable students to explore digital solutions to the challenges. The relevant digital technology curriculum descriptors are included below.

1. Forbes, A., Chandra, V., Pfeiffer, L., & Sheffield, R. (2021). *STEM education in the primary school: a teacher's toolkit*. Cambridge University Press.

Connections to the Australian Curriculum
(Digitial Technologies, Science, Design and Technologies, and Mathematics)

Subject	Content Descriptions	Activity							
		1	2	3	4	5	6	7	8
DIGITAL TECHNOLOGIES	define problems with given or co-developed design criteria and by creating user stories (AC9TDI6P01)	×	×	×	×	×	×	×	×
	design algorithms involving multiple alternatives (branching) and iteration (AC9TDI6P02)							×	
	design a user interface for a digital system (AC9TDI6P03)							×	
	implement algorithms as visual programs involving control structures, variables and input (AC9TDI6P05)							×	
	evaluate existing and student solutions against the design criteria and user stories and their broader community impact (AC9TDI6P06)	×	×	×	×	×	×	×	×
	select and use appropriate digital tools effectively to create, locate and communicate content, applying common conventions (AC9TDI6P07)	×	×		×	×	×	×	×

Connections to the Australian Curriculum
(Digitial Technologies, Science, Design and Technologies, and Mathematics)

Subject	Content Descriptions	Activity 1	2	3	4	5	6	7	8
SCIENCE	examine how particular structural features and behaviours of living things enable their survival in specific habitats (AC9S5U01)	×						×	×
	describe how weathering, erosion, transportation and deposition cause slow or rapid change to Earth's surface (AC9S5U02)	×							
	identify sources of light, recognise that light travels in a straight path and describe how shadows are formed and light can be reflected and refracted (AC9S5U03)		×						
	explain observable properties of solids, liquids and gases by modelling the motion and arrangement of particles (AC9S5U04)			×					
	investigate how scientific knowledge is used by individuals and communities to identify problems, consider responses and make decisions (AC9S5H02)	×	×	×	×	×	×	×	×
	pose investigable questions to identify patterns and test relationships and make reasoned predictions (AC9S5I01)	×	×	×	×	×	×	×	×
	plan and conduct repeatable investigations to answer questions, including, as appropriate, deciding the variables to be changed, measured and controlled in fair tests; describing potential risks; planning for the safe use of equipment and materials; and identifying required permissions to conduct investigations on Country/Place (AC9S5I02)	×	×	×	×	×	×	×	×
	use equipment to observe, measure and record data with reasonable precision, using digital tools as appropriate (AC9S5I03)	×	×	×	×	×	×	×	×
	construct and use appropriate representations, including tables, graphs and visual or physical models, to organise and process data and information and describe patterns, trends and relationships (AC9S5I04)	×	×	×	×	×	×	×	×
	compare methods and findings with those of others, recognise possible sources of error, pose questions for further investigation and select evidence to draw reasoned conclusions (AC9S5I05)	×	×	×	×	×	×	×	×
	write and create texts to communicate ideas and findings for specific purposes and audiences, including selection of language features, using digital tools as appropriate (AC9S5I06)	×	×	×	×	×	×	×	×
	investigate the physical conditions of a habitat and analyse how the growth and survival of living things is affected by changing physical conditions (AC9S6U01)	×						×	×
	investigate the transfer and transformation of energy in electrical circuits, including the role of circuit components, insulators and conductors (AC9S6U03)				×				
	compare reversible changes, including dissolving and changes of state, and irreversible changes, including cooking and rusting that produce new substances (AC9S6U04)			×					
	investigate how scientific knowledge is used by individuals and communities to identify problems, consider responses and make decisions (AC9S6H02)	×	×	×	×	×	×	×	×
	pose investigable questions to identify patterns and test relationships and make reasoned predictions (AC9S6I01)	×	×	×	×	×	×	×	×
	plan and conduct repeatable investigations to answer questions including, as appropriate, deciding the variables to be changed, measured and controlled in fair tests; describing potential risks; planning for the safe use of equipment and materials; and identifying required permissions to conduct investigations on Country/Place (AC9S6I02)	×	×	×	×	×	×	×	×
	use equipment to observe, measure and record data with reasonable precision, using digital tools as appropriate (AC9S6I03)	×	×	×	×	×	×	×	×
	construct and use appropriate representations, including tables, graphs and visual or physical models, to organise and process data and information and describe patterns, trends and relationships (AC9S6I04)	×	×	×	×	×	×	×	×
	compare methods and findings with those of others, recognise possible sources of error, pose questions for further investigation and select evidence to draw reasoned conclusions (AC9S6I05)	×	×	×	×	×	×	×	×
	write and create texts to communicate ideas and findings for specific purposes and audiences, including selection of language features, using digital tools as appropriate (AC9S6I06)	×	×	×	×	×	×	×	×

Connections to the Australian Curriculum
(Digitial Technologies, Science, Design and Technologies, and Mathematics)

Subject	Content Descriptions	Activity 1	2	3	4	5	6	7	8
DESIGN AND TECHNOLOGIES	explain how people in design and technologies occupations consider competing factors including sustainability in the design of products, services and environments (AC9TDE6K01)	×	×	×	×	×	×	×	×
	explain how electrical energy can be transformedinto movement, sound or light in a product or system (AC9TDE6K02)				×				
	explain how and why food and fibre are produced in managed environments (AC9TDE6K03)	×	×						
	explain how characteristics and properties of materials, systems, components, tools and equipment affect their use when producing designed solutions (AC9TDE6K05)	×	×	×	×	×	×	×	×
	investigate needs or opportunities for designing, and the materials, components, tools, equipment and processes needed to create designed solutions (AC9TDE6P01)	×	×	×	×	×	×	×	×
	generate, iterate and communicate design ideas, decisions and processes using technical terms and graphical representation techniques, including using digital tools (AC9TDE6P02)	×	×	×	×	×	×	×	×
	select and use suitable materials, components, tools, equipment and techniques to safely make designed solutions (AC9TDE6P03)	×	×	×	×	×	×	×	×
	negotiate design criteria including sustainability to evaluate design ideas, processes and solutions (AC9TDE6P04)	×	×	×	×	×	×	×	×
	develop project plans that include consideration of resources to individually and collaboratively make designed solutions (AC9TDE6P05)	×	×	×	×	×	×	×	×
MATHEMATICS	choose appropriate metric units when measuring the length, mass and capacity of objects; use smaller units or a combination of units to obtain a more accurate measure (AC9M5M01)								×
	solve practical problems involving the perimeter and area of regular and irregular shapes using appropriate metric units (AC9M5M02)	×	×	×	×	×	×		×
	estimate, construct and measure angles in degrees, using appropriate tools including a protractor, and relate these measures to angle names (AC9M5M04)								×
	construct a grid coordinate system that uses coordinates to locate positions within a space; use coordinates and directional language to describe position and movement (AC9M5SP02)				×		×		
	acquire, validate and represent data for nominal and ordinal categorical and discrete numerical variables to address a question of interest or purpose using software including spreadsheets; discuss and report on data distributions in terms of highest frequency (mode) and shape, in the context of the data (AC9M5ST01)				×		×		
	interpret line graphs representing change over time; discuss the relationships that are represented and conclusions that can be made (AC9M5ST02)				×		×		
	establish the formula for the area of a rectangle and use it to solve practical problems (AC9M6M02)				×				×
	compare the parallel cross-sections of objects and recognise their relationships to right prisms (AC9M6SP01)								×
	use mathematical modelling to solve practical problems, involving rational numbers and percentages, including in financial contexts; formulate the problems, choosing operations and efficient calculation strategies, and using digital tools where appropriate; interpret and communicate solutions in terms of the situation, justifying the choices made (AC9M6N09)	×	×	×	×	×	×	×	×
	locate points in the 4 quadrants of a Cartesian plane; describe changes to the coordinates when a point is moved to a different position in the plane (AC9M6SP02)					×		×	
	create and use algorithms involving a sequence of steps and decisions that use rules to generate sets of numbers; identify, interpret and explain emerging patterns (AC9M6A03)							×	
	plan and conduct statistical investigations by posing and refining questions or identifying a problem and collecting relevant data; analyse and interpret the data and communicate findings within the context of the investigation (AC9M6ST03)							×	

Challenge Activities

Notes for the student

In this workbook, you will engage in eight activities. In each activity, you will use science, technology, engineering and mathematics, or STEM, to develop solutions to real-world challenges.

What will I do in the Challenge activities?

In each challenge activity you will follow these steps:

Step 1. Ask
What is the challenge?

Step 2. Imagine
How can I tackle the challenge?

Step 3. Plan
How can I plan my idea?

Step 4. Create
What will my idea look like?

Step 5. Improve
How can I make my idea better?

Then, at the end, you will **reflect** on the activity and its outcomes to see what you have learnt.

The use of digital technologies is encouraged. Throughout the challenges you will get the opportunity to use digital technology tools to research, investigate, create, and share your knowledge and understanding. These experiences will also enable you to think deeper about digital safety. Ask your teacher/adult if in doubt.

Pages 105–108 at the back of this workbook provide some additional space for notes, and practice grids for you to test your ideas at the planning stage. Use the blank page if you wish to record any interesting things as you go along.

Now, let's have a look at the eight Challenge activities. Each activity is based on a different United Nations Sustainable Development Goal...

Challenge 1
— A permaculture garden
Goal: Zero Hunger

Challenge 2
— A SunSmart hat
Goal: Good Health and Well-being

Challenge 3
— A water-filtration device
Goal: Clean Water and Sanitation

Challenge 4
— A solar solution for lighting spaces
Goal: Affordable and Clean Energy

Challenge 5
— A kit shelter for refugees
Goal: Sustainable Cities and Communities

Challenge 6
— A toy that moves made from waste
Goal: Climate Action

Challenge 7
— A video game on plastic-waste removal
Goal: Life below Water

Challenge 8
— A habitat for giraffes
Goal: Life on Land

My STEM Workbook 3 – Understanding Science, Technology, Engineering and Mathematics through design-process activities — Vinesh Chandra and Basil Slynko — ISBN: 978-0-6484052-2-1

The United Nations Sustainable Development Goal 2
Zero Hunger

is about ending hunger, achieving food security, improving nutrition and promoting sustainable agriculture. One of the ways to do this is by growing your own food. Growing food means growing vegetables, fruits, herbs and spices.

Key words in this activity:
- companion plants
- compostable
- cyclical system
- deposition
- ecosystem
- erosion
- organisms
- permaculture
- sustainable
- system
- water capture
- weathering

Challenge 1. A permaculture garden

A permaculture garden

An agronomist has challenged your class to plan and build a permaculture garden. The local community wants to introduce **permaculture** gardening. Your challenge is to locate a site and then design and build a permaculture garden. You have to grow a selection of plants that support the needs of the community.

An agronomist is someone whose job it is to grow crops and also take care of the soil.

1. Ask

Permaculture is a garden that can essentially take care of itself. Plants that grow in a permaculture garden are part of a sustainable **system**. The design of this system is such that the plants fit in harmony with people and the **ecosystem**.

A system is an organised way of doing things.

The basic parts of a system

*A **sustainable** system is one that can take care of itself, with little or no help from people.*

*An **ecosystem** is a collection of living and non-living **organisms** that interact with each other in a specific area.*

In a system, different parts connect with each other to produce an outcome or complete a task.

Permaculture is also known as **perma**nent agri**culture**

The three principles of permaculture are:
caring for the Earth
caring for people
sharing fairly with others and the Earth.

Permaculture gardens are designed to regenerate and sustain the garden. Permaculture gardens use ground cover, which protects and nourishes the soil; and attracts birds, pollinators/bees and insects to repel pests. Ground cover also limits **weathering** and erosion of earthen materials as well as the deposition of sediment by wind and water. Without ground cover, earthen materials can be removed, which can change the landscape and expose the roots of plants to the elements.

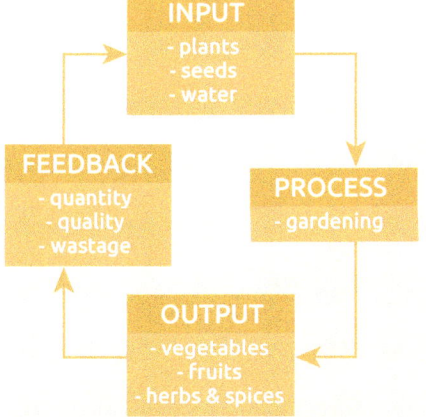

A permaculture system

Challenge 1. A permaculture garden

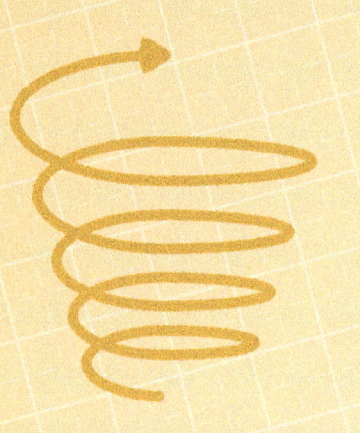

Permaculture is a cyclical system

The soil is improved with compost, straw, green waste or farmyard waste. Plants are carefully chosen so that they are good companions of each other. Plants are good companions when they help each other grow. For example, marigold is a good companion of tomato plants because it keeps away some pests. Sweet corn is a good companion of climbing beans because it allows them to grow on it.

Companion plants are grouped near one another and sometimes stacked vertically. The advantage is that smaller plants are protected by larger plants and there are water savings. However, plants that dislike growing with others are planted apart. For example tomatoes and cabbage, beans and onions, or carrots and celery.

Permaculture is not linear as there is no clear beginning or end. It is cyclical – that is, it is a **cyclical system**. The output generally increases every year.

Q1a. Why do we need to consider permaculture as an option for gardening? Explain in your own words.

_____
_____
_____
_____
_____
_____
_____
_____

Research

Do some research in your library and on the Internet to find answers to the questions that follow. This may also help you to think through some of the other questions in this activity.

Here are some URLs to get your research underway:

1. YouTube video: https://youtu.be/jxP4HlFIpQw?t=49
   Video title search terms: What is permaculture?

2. YouTube video: https://youtu.be/SqBdKsHZP6w
   Video title search terms: What is permaculture?

3. Website: https://www.permaculturenorthernbeaches.org.au/permaculture-for-kids
   Webpage title search terms: Permaculture for kids

4. YouTube video: https://www.youtube.com/watch?v=tgophFI451Y
   Video title: Best Ways to Collect Rainwater Search terms: Free water / rain water

5. Website: https://seelibrary.org/   Webpage Title: SEE library
   Look up Science and Technology category.

Challenge 1. A permaculture garden

Q1b. When was permaculture conceived and developed?

_____

_____

_____

Q1c. By whom was permaculture conceived and developed?

Bill _____ and David _____

Q1d. What is water capture/harvesting?

_____

_____

_____

_____

Some things to consider when designing a permaculture garden are:
- *the natural features of the site*
- *whether the ground is flat or sloping*
- *that there is plenty of sun or shade*
- *the height of trees*
- *any native plants*
- *companion plants*
- *raised beds*
- **compostable materials**
- *garden design*
- **water capture**/*harvesting*
- *the orientation of the sun*
- *erosion and deposition.*

The natural materials of permaculture gardens should not be removed. Removing materials can change the landscape and expose the roots of plants to the elements.

_____

Q1e. How is soil **erosion** controlled in permaculture? List two ways.

1. _____

2. _____

Q1f. How is sediment **deposition** controlled in permaculture?

_____

_____

_____

_____

Challenge 1. A permaculture garden

2. Imagine

Q2a. What vegetables do you want to grow?
Use the table to list three vegetables, then list one companion plant for each vegetable.

	Vegetable	Companion plant
1		
2		
3		

Q2b. What fruit trees and vines could you plant to match the needs of the community? List three fruit trees and three vines in the table.

	Fruit tree	Vine
1		
2		
3		

Compostable materials such as straw, green waste and farmyard waste are commonly used to nourish and protect the soil.

Q2c. List three other compostable materials that you could use in your permaculture garden.

1. _____ 2. _____ 3. _____

Q2d. What water capture techniques could you use in your permaculture garden? List three.

1. _____

2. _____

3. _____

Garden design in permaculture is based around 'zones'.

Q2e. What are zones?

_____

_____

Q2f. Why are zones significant?

_____

_____

3. Plan

Q3a. Use the grid to draw a rough sketch of the site and your initial idea for a permaculture garden.

Show your sketch to your classmates and get their feedback.

Some things to consider are:
- features of the site
- water capture/harvesting techniques
- path of the sun
- companion plants
- erosion prevention measure.

Q3b. What did your classmates think? Write their comments here:

_____
_____
_____
_____
_____
_____
_____
_____
_____
_____

Consider your classmates' feedback.

Q3c. Draw your final design here.
Remember to:
- label the main features such as plants and irrigation system
- add any other information such as plant details and zones
- also show dimensions.

Show your final design to your teacher and get their comments and approval.

Q3d. Use the table below to list the materials (including plants), their purpose and the quantities you will need for your permaculture garden.

Material	Purpose	Quantity

Procedure

Q3e. What is the procedure for preparing your permaculture garden? Draw a line from each step number to the correct task.

Step	Task
1	Design the garden
2	Get the plants
3	Install watering system
4	Mark out the garden
5	Mark out the planting
6	Observe the site
7	Plant the plants
8	Add compost
9	Select a site

Challenge 1. A permaculture garden

Tools & Equipment

Safety is important to prevent accidents and injuries when gardening. Tools and equipment can be dangerous if not used correctly. You should always wear safety gear to protect yourself. The other term for safety gear is "Personal Protective Equipment" or "PPE".

Here are some gardening tools that may be needed for a permaculture garden.

a. Hoe; b. Hand fork; c. Rake; d. Shovel; e. Pruning saw; f. Secateurs.

Q3f. Select three gardening tools that you are most likely to use in your permaculture garden. Write how each tool could injure you and what safety gear you should wear to prevent the injury.

1 Gardening tool:_____

How can I get injured? _____

_____

What safety gear should I wear?_____

_____

2 Gardening tool:_____

How can I get injured? _____

_____

What safety gear should I wear?_____

_____

3 Gardening tool:_____

How can I get injured? _____

_____

What safety gear should I wear?_____

_____

4. Create

Making a permaculture garden

Seek feedback from your teacher while you are creating your garden.

You need to:
- prepare your site
- gather the plants for your permaculture garden
- gather the tools and equipment you need
- install the watering system
- follow your procedure to make your garden
- use PPE and follow the safety rules.

Permaculture is a gardening system with many parts.

Q4a. Draw a section of your established garden and use arrows to show the zones and how the different parts interconnect with each other.

Q4b. Making your permaculture garden may not go as planned. For each step in the procedure that caused problems, explain how you dealt with it. List the step(s) and your course of action.

_____
_____
_____
_____
_____
_____
_____
_____
_____
_____
_____
_____
_____
_____
_____
_____
_____
_____
_____
_____

5. Improve

Present your permaculture-garden idea to the class. Convince them that your idea aligns with the UN's goal and has addressed the challenge for this activity.

Ask your classmates these questions:
- What are some of the good points about my permaculture-garden idea?
- What are some of the weak points about my permaculture-garden idea?

Q5a. Write the most useful answers from your classmates in the table.

Good points

Weak points

Consider your classmates' answers

Q5b. What could you do to improve your permaculture-garden idea?

_____
_____
_____

Now it's time to reflect...

What have you learnt in this activity? *List three things.*
Hint: Look back at the key words.

1. _____

2. _____

3. _____

What would you like to know more about?

_____
_____
_____

 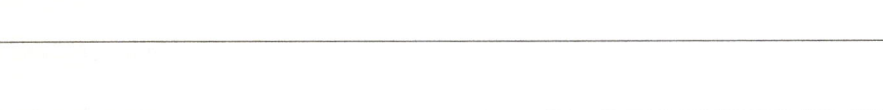

Try the quiz on Kahoot. See your teacher.

How can you set up a mini-permaculture garden where you live? You may like to use a computer application like Canva to illustrate your idea?

Need more space? Use pages 105–108.

Challenge 1. A permaculture garden

The United Nations Sustainable Development Goal 3

Good Health and Well-being

is about ensuring people lead healthy lives and promoting the well-being of everyone. Good health and well-being is about looking after ourselves. Skin cancer is a problem in our communities. Harmful **rays** from the sun are a major cause of this condition. We can protect ourselves from these harmful rays by being SunSmart.

Key words in this activity:

- area
- electromagnetic radiation
- infrared radiation (IR)
- materials
- rays
- shadow
- skin cancer
- SunSmart
- UV light
- visible light

A SunSmart hat

Challenge 2

A fashion designer has challenged students to design and make models of SunSmart hats that reinforce the 'slap on a hat' theme. Your class is taking part in this event. Your teacher would like you to make a model of a SunSmart hat to suit a child of your age. Outstanding examples will be displayed in the library.

Information: A fashion designer is a person who designs clothes and accessories. They also advise how to make the item and check its quality.

1. Ask

Q1a. Why do we need to be SunSmart?

_____

_____

_____

Q1b. How does a hat make us SunSmart?

_____

_____

_____

Research

Do some research in your library and on the Internet to find answers to the questions that follow. This may also help you to think through some of the other questions in this activity.

Here are some URLs to get your research underway:

1. Website: https://www.cancer.org.au/cancer-information/causes-and-prevention/sun-safety/campaigns-and-events/national-skin-cancer-action-week
   Webpage Title: Combatting Australia's national cancer

2. Website: https://www.cancer.org.au/about-us/how-we-help/prevention/stories/turning-the-tide-of-skin-cancer   Webpage Title: Turning the tide on skin cancer

3. Website: https://www.sunsmart.com.au/downloads/schools-early-childhood/vels/make-a-sunsmart-hat-sequence.pdf   Webpage Title: Make a SunSmart hat

4. Website: https://seelibrary.org/   Webpage Title: SEE library
   Look up the "Science and Technology" category.

Q1c. What causes **skin cancer**?

_____

_____

_____

_____

Q1d. What does "Slip, Slop, Slap, Seek and Slide" mean? How do these actions make us **SunSmart**?

Action	What does it mean?	How does it make us SunSmart?
Slip		
Slop		
Slap		
Seek		
Slide		

Challenge 2. A SunSmart hat

2. Imagine

The Sun is a giant ball of hot gas that is constantly emitting energy in all directions.

The Sun emits energy in the form of **electromagnetic radiation**, including **visible light**, ultraviolet radiation (UV), and **infrared radiation (IR)**. The light you see every day is called visible light. The heat you feel is due to infrared light.

However, you cannot see or feel UV light. This is the most dangerous of the three.

The Sun's energy reaches the Earth through a process called radiation. This radiation travels through space at the speed of light and reaches the Earth about 8 minutes after it is emitted by the Sun.

> **Information**
>
> The Cancer Council recommends that we "Slap on a broad-brimmed hat that shades your face, head, neck and ears."

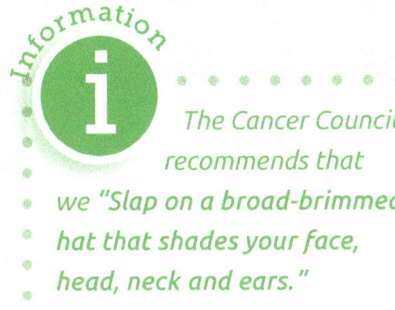

Q2a. When electromagnetic radiation reaches the Earth, it generally travels in a straight line. How do we know this? Explain your answer and include a diagram. *Hint: think about shadows.*

Q2b. UV light is considered to be the most dangerous because it can cause skin cancer. Why is UV light so dangerous?

Q2c. What can we do to protect ourselves from skin cancer?

Q2d. A broad-brimmed hat and a cap are shown above. Why is a broad-brimmed hat better for sun protection than a cap?

Challenge 2. A SunSmart hat

Q2e. Shelley invites Mary to join her for a game of handball. It is an overcast day and Shelley suggests that there is no need to wear a hat. Do you agree with Shelley's advice? Explain your answer.

_____
_____
_____

Q2f. There are a number of factors which you need to consider when you design and make a hat. One of the most import factors would be your ability to make the hat. Some other factors relate to the design of the hat and include appearance, comfort, colour, cost, **materials**, and Sun protection. Rank these factors (from most to least important) and explain their importance.

	Factor	Why is this factor important?
1		
2		
3		
4		
5		
6		

Q2g. The following are some examples of materials that you can use to make a hat. What is an advantage and disadvantage of using the material?

Material	Advantage	Disadvantage
Cardboard		
Paper		
Plastic		
Fabric		
Fur		
Straw		
Leaves		

3. Plan

Q3a. Use the grid to draw a rough sketch of your SunSmart hat.

Show your sketch to your classmates and get their feedback.

Q3b. What did your classmates think? Write their comments here:

_____
_____
_____
_____
_____
_____
_____
_____
_____
_____
_____
_____
_____
_____

Consider your classmates' feedback.

Q3c. Review your initial sketch. Now, draw the final sketch of your SunSmart hat. Some things you should consider in the final design are:
- dimensions
- shape.

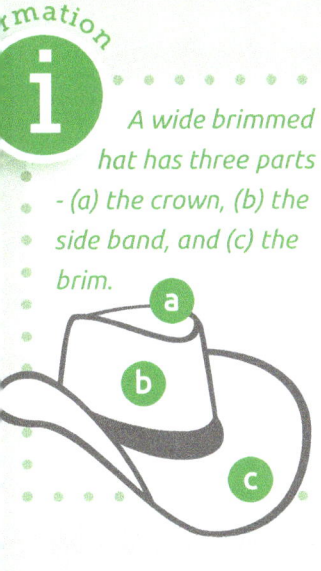

A wide brimmed hat has three parts - (a) the crown, (b) the side band, and (c) the brim.

Q3d. Estimate how much material you would need to make your SunSmart hat. In the table below sketch the parts, list the dimensions and work out the **area** for each part.

Part	Sketch	
Crown		
	Dimensions	Area
Side band		
	Dimensions	Area
Brim		
	Dimensions	Area

Q3e. What materials will you use to make your SunSmart hat? Use the table below to list the materials, their main purpose and the quantities you will need.

Material	Main purpose	Quantity

Procedure

Q3f. Write the steps for making the model of your SunSmart hat.

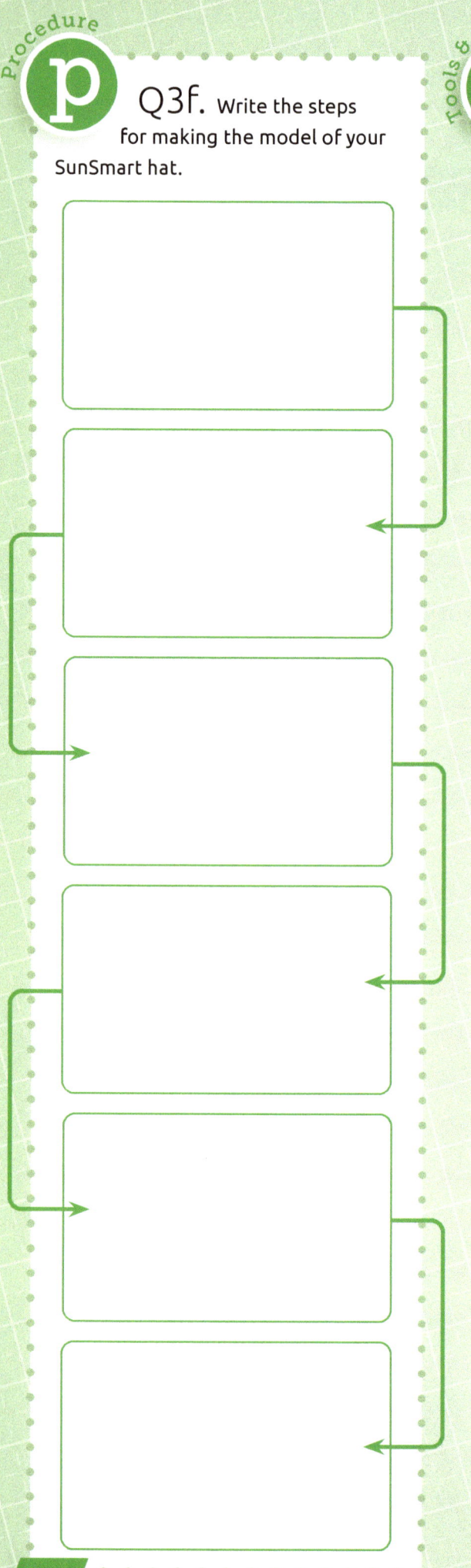

Tools & Equipment

Safety is important when you are outdoors. Safety is also important when you are making models. Sometimes you may need safety gear. The other term for safety gear is "Personal Protective Equipment" or "PPE". For every activity you undertake, you need to understand the safety rules and the possible consequences of not following them.

Q3g. What are the main safety rules that you need to follow when making a hat? What are the consequences if they are not followed? List them in the table. *Remember, you may also need to go out in the sun to test your hat.*

Safety rule	Consequences if the rule is not followed

Challenge 2. A SunSmart hat

4. Create

Making your SunSmart hat model

Seek feedback from your teacher while you are making your model.

You need to:
- gather the materials
- gather the equipment
- follow your procedure to build your hat model
- follow the safety rules.

Q4a. Making your hat model may not go as planned. For each step in the procedure that caused problems, explain how you dealt with it. List the step(s) and your course of action.

_____
_____
_____
_____
_____
_____
_____
_____
_____
_____
_____
_____
_____
_____
_____
_____
_____

Test

Q4b. Does your model cover your face, head, neck and ears at different times of the day? Do a "SunSmart Hat Test". You can do this with a partner. At different times of the day, put on your hat and step out in the sun. Your partner can tell you if your hat is providing shade to your face, head, neck and ears. Record a tick or a cross in the table below. *Ensure that you are standing in the same spot each time.*

Time	Face	Head	Neck	Ears
e.g. 9am	✓	✓	✗	✓

Q4c. Why was standing in the same spot important?

_____
_____
_____
_____
_____

Q4d. How can you explain your results?

_____
_____
_____
_____
_____
_____
_____
_____
_____
_____
_____

5. Improve

Present your SunSmart hat model to the class. Convince them that your idea aligns with the UN's goal and has addressed the challenge for this activity.

Ask your classmates these questions:
- What are some of the good points about my SunSmart hat?
- What are some of the weak points about my SunSmart hat?

Q5a. Write the most useful answers from your classmates in the table.

Good points

Weak points

Challenge 2. A SunSmart hat

Consider your classmates' answers

Q5b. What could you do to improve your SunSmart hat idea?

_____
_____
_____

Now it's time to reflect...

What have you learnt in this activity? *List three things.*
Hint: Look back at the key words.

1. _____
2. _____
3. _____

What would you like to know more about?

_____
_____
_____
_____
_____
_____

q *Try the quiz on Kahoot. See your teacher.*

Curly Question

The Cancer Council of Australia has developed the SunSmart App. Read more about it on the Cancer Council of Australia website (https://www.cancer.org.au/cancer-information/causes-and-prevention/sun-safety/be-sunsmart/sunsmart-app). Do the "How SunSmart Are You" Quiz on this webpage.

How can you create your own SunSmart app? You can use an application like Microsoft PowerPoint to create a prototype. (Hint: Search the Internet for ideas)

The United Nations Sustainable Development Goal 6
Clean Water and Sanitation

is about ensuring that there is clean water for everyone. For many people in the world, access to clean water that flows from taps is a problem. One of the targets of Goal 6 is to find ways for everyone to have access to clean water.

Key words in this activity:
- 3D
- charcoal
- filter
- filtration
- insoluble
- liquid
- sediments
- solid
- soluble
- solution
- suspension

Challenge 3. A water-filtration device

A water-filtration device

A water engineer has challenged your class to design and create models of water-filtration devices that can be used in developing countries. Water from your water-filtration device should be useful for cleaning and washing. Your teacher would like you to create your model using materials that are easily accessible. Your model should have the capacity to **filter** at least 200 millilitres (mL) of water.

> **Information**
> *A water engineer is a professional who specialises in the design and management of water and sanitation systems and processes. They look after dams, water supply, wastewater and stormwater.*

1. Ask

Q1a. Why do we need clean water?

_____

_____

_____

Q1b. Most of the water that flows through our taps comes from rivers and dams. What are three main differences between tap water and water in rivers and dams?

1 _____

_____

2 _____

_____

3 _____

_____

Q1c. What is the risk of drinking untreated water from rivers and dams?

_____

_____

_____

_____

Do some research in your library and on the Internet to find answers to the questions that follow. This may also help you to think through some of the other questions in this activity.

Here are some URLs to get your research underway:

1. YouTube Video: https://www.youtube.com/watch?v=ICYNtiU7r6I
   Video title search terms: How to make a water filter with sand and **charcoal**

2. YouTube Video: https://www.youtube.com/watch?v=IH-2HyTpmC0
   Video title search terms: How to make a simple water filter project for school

3. Website: https://kids.nationalgeographic.com/books/article/water-wonders
   Webpage Title: National Geographic Kids – Make a Water Filter

4. Website: https://seelibrary.org/   Webpage Title: SEE library
   Look up the "Science and Technology" category.

Q1d. What information did you find about water filters from your research? Here is a PMI table. The acronym PMI stands for "Pluses, Minuses, Interesting". It allows you to list what was good (pluses), what was not so good (minuses) and what you found interesting in your research. List two points in each column.

Pluses (+)	Minuses (-)	Interesting (?)

2. Imagine

Some water from a muddy pond was placed in a bottle. The images show the changes that occurred in the bottle over a few days.

Start | After 1 day | After 2 days

Q2a. What was happening in the bottles? Use the words **solid**, **liquid**, **soluble**, particles, **sediments**, **insoluble**, **solution**, and **suspension** in your answer.

Start: _____

_____

_____

_____

After 1 day: _____

_____

_____

_____

After 2 days: _____

_____

_____

_____

Q2b. After three days, the water in the bottle appeared clear with some **sediments** at the bottom. But clear water is not always fit for drinking. Why not?

_____

_____

_____

_____

_____

_____

Q2c. What is water **filtration**?

_____

_____

_____

_____

Q2d. How does filtration remove sediments from dirty water?

_____

_____

_____

_____

Q2e. What are the key features of a water-filtration device? Explain your answer with a diagram.

3. Plan

Q3a. Use the grid to draw a rough sketch of your water-filtration device. Your sketch can be a side view or drawn in **3D**.

Show your sketch to your classmates and get their feedback.

Q3b. What did your classmates think? Write their comments here:

_____

_____

_____

_____

_____

_____

_____

_____

_____

_____

_____

_____

Consider your classmates' feedback.

Q3c. Review your initial sketch. Now, draw the final sketch of your water-filtration device. **Do not forget to label your device.**

Q3d. What materials will you use to make your water-filtration device model? Use the table below to list the materials, their purpose and the quantities you will need.

Material	Purpose	Quantity

Safety is important to prevent accidents and injuries. Using tools and equipment can be dangerous. You may need to wear safety gear to protect yourself when making a purification device. The other term for safety gear is "Personal Protective Equipment" or "PPE". For every activity you undertake, you need to understand the safety rules and the possible consequences of not following them.

Q3e. What are the main safety rules that you need to follow when making a filtration device? What are the consequences if they are not followed? List them in the table.

Safety rule	Consequences if the rule is not followed

Some examples of PPE: face mask, protective gloves, protective glasses and an apron.

Q3f. What procedure will you follow to create your water-filtration device? Write the steps here.

4. Create

Making a water-filtration device

Seek feedback from your teacher while you are making your device.

You need to:
- gather materials
- gather the equipment
- follow your procedure
- follow the safety rules.

Q4a. Making your water-filtration device may not go as planned. For each step in the procedure that caused problems, explain how you dealt with it. List the step(s) and your course of action.

_____
_____
_____
_____
_____
_____
_____
_____
_____
_____
_____
_____
_____
_____
_____

Test your device. One of the ways to do this is by recording the volume of water that is filtered over time. To get this data, you will need a measuring cylinder and a stopwatch.

Q4b. Record and graph the data. You can also use an application like Microsoft Excel to do this question.

Time (min)						
Volume of water filtered (ml)						

Q4c. What is the graph showing you?

_____
_____
_____
_____
_____
_____
_____
_____
_____
_____
_____
_____
_____
_____
_____

Graph Title: _____

Label (Y axis): _____

Label (X axis): _____

Q4d. Determine the average rate at which water flows from your filtration device. In this case, the rate can be determined by the volume of water that is filtered over time. Write your answers in the space below.

Volume of water filtered = _____ millilitres (ml)

Time taken = _____ minutes (min.)

$$\text{Rate} = \frac{\text{Volume of water filtered}}{\text{Time taken}} = \frac{\quad\quad}{\quad\quad} = \_\_\_\_ \text{ ml/min.}$$

Challenge 3. A water-filtration device

Access to safe drinking water
What share of people have access to safe drinking water?

SDG Target 6.1 is to: "achieve universal and equitable access to safe and affordable drinking water for all" by 2030.

In 2020, almost three-quarters (74%) of the world population had access to a safely managed water source. One-in-four people do not have access to safe drinking water.

In the chart we see the breakdown of drinking water access globally, and across regions and income groups. We see that in countries at the lowest incomes, less than one-third of the population have safe water. Most live in Sub-Saharan Africa.

The world has made progress in the last five years. Unfortunately, this has been very slow. In 2015 (at the start of the SDGs) only 70% of the global population had safe drinking water. That means we've seen an increase of four percentage points over five years.

This is obviously far too slow to reach universal access by 2030. If progress continues at these rates, we would only reach 82% by 2030. If we're to meet our target we need to see rates of progress more than triple (increase 3.2-fold) for the coming decade.[1]

1. To reach universal access within 10 years (the current gap is 26%) we need to see a 13 percentage point increase every five years. That's 3.25 times higher than the 4 percentage point increase we've seen in the last five years.

*Data source: WHO/UNICEF Joint Monitoring Programme for Water Supply, Sanitation and Hygiene (JMP) – OurWorldInData.org/water-access

Share of the population using safely managed drinking water (2022)*

(A safely managed drinking water service is defined as one located on premises, available when needed and free from contamination.)

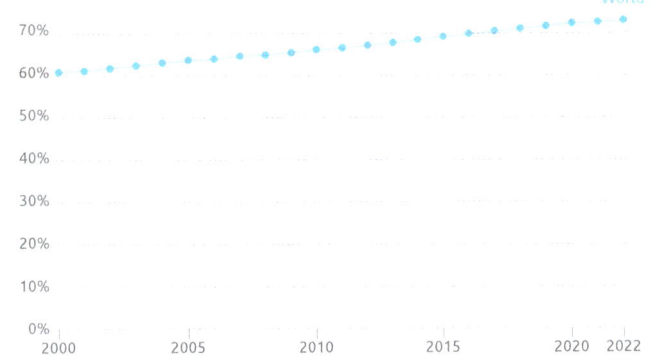

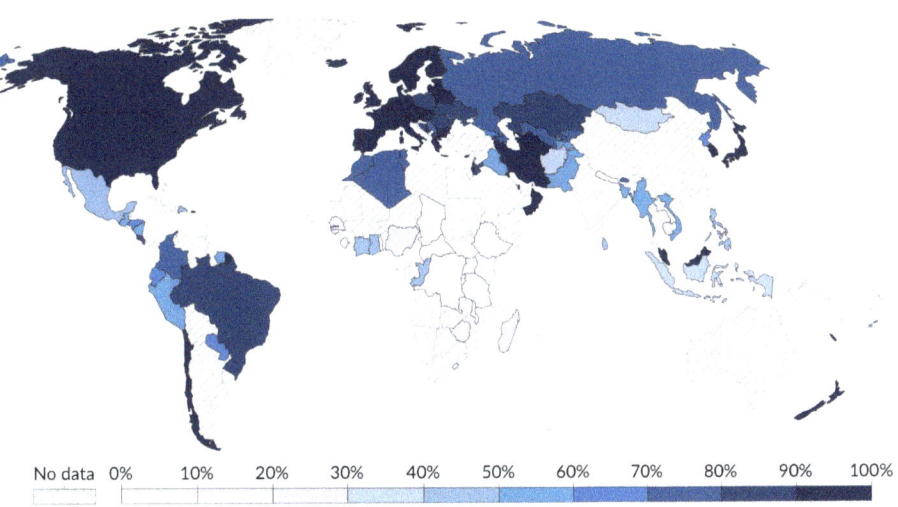

Share of the population with access to drinking water facilities (2022)*

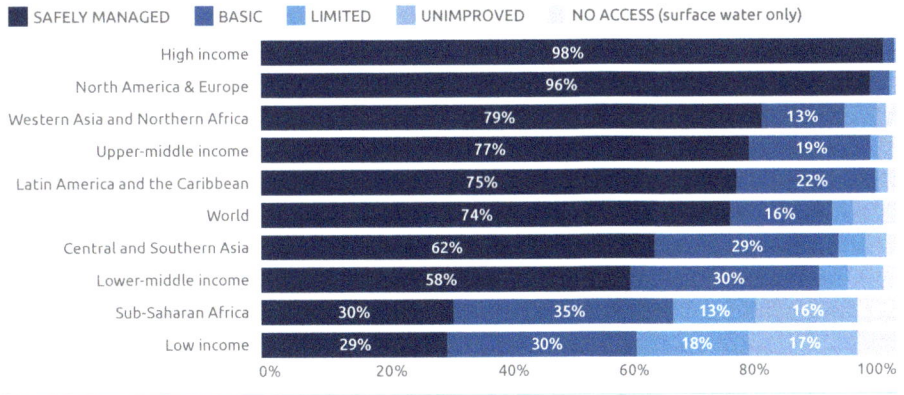

Article source: Hannah Ritchie, Fiona Spooner and Max Roser (2019) - "Clean Water" Published online at OurWorldInData.org. Retrieved from: 'https://ourworldindata.org/clean-water' [Online Resource]

Challenge 3. A water-filtration device

5. Improve

Present your water-filtration device idea to the class. Convince them that your idea aligns with the UN's goal and has addressed the challenge for this activity.

Ask your classmates these questions:
- *What are some of the good points about my water-filtration device?*
- *What are some of the weak points about my water-filtration device?*

Q5a. Write the most useful answers from your classmates in the table.

Good points

Weak points

Consider your classmates' answers

Q5b. What could you do to improve your water-filtration device idea?

_____

_____

_____

Now it's time to reflect...

What have you learnt in this activity? *List three things.*
Hint: Look back at the key words.

1. _____

2. _____

3. _____

What would you like to know more about?

_____

_____

_____

 *Try the quiz on Kahoot. See your teacher.*

 How do water-purification tablets work? Can they be used to give more people access to clean water?

Need more space? Use pages 105–108.

The United Nations Sustainable Development Goal 7

Affordable and Clean Energy

is about ensuring access to affordable, reliable, sustainable and modern energy for all. We need to use sustainable energy sources such as solar, wind and water to power lights, air conditioners, heaters, washing machines and other appliances. These sources are much cleaner and less likely to impact climate change. One of the targets of Goal 7 is to explore ways for more renewable energy sources to be used to meet human needs.

Key words in this activity:
- circuit
- solar energy
- solar panel
- switch
- three dimensional
- variables

Challenge 4. A solar solution for lighting spaces

A solar solution for lighting spaces

An electrician has challenged your class to light a garden shed using clean energy. Many garden sheds do not have lights. She would like your class to design and create a model of a circuit that can be used to light a garden shed using a solar panel. The shed will also need to have two lights, each one with its own switch. Your teacher would like you to tackle this challenge and show how your circuit will work using a small cardbox box.

> **Information**
> An electrician is a person who is qualified to undertake tasks associated with electricity, such as house wiring.

To help you think through the challenge, a 3D view and a floor plan of the garden shed are shown below.

3D view of the garden shed

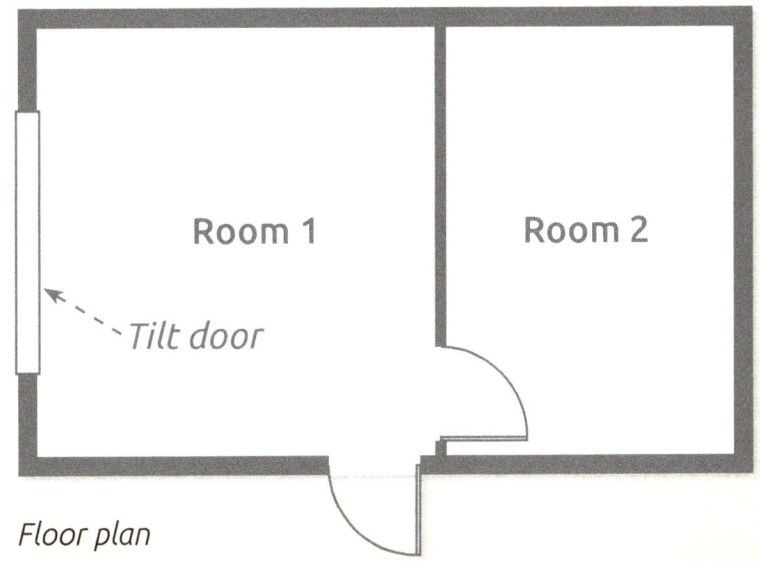

Floor plan

1. Ask

Q1a. Why do we need clean energy?

Q1b. Why do we need affordable energy?

Q1c. When you turn on a switch in your bedroom, a light comes on. It is very likely that only the light in your room turns on. How does this happen?

Q1d. Why should a solar option be explored for lighting your garden shed?

r Research

Do some research in your library and on the Internet to find answers to the questions that follow. This may also help you to think through some of the other questions in this activity.

Here are some URLs to get your research underway:

1. Website: https://theconversation.com/curious-kids-how-do-solar-panels-work-123515   Webpage Title: How do solar panels work conversation
2. Website: https://www.renewableenergyworld.com/solar/solar-energy-for-kids/#gref
   Webpage Title: Solar energy for kids
3. Website: https://seelibrary.org/   Webpage Title: SEE library
   Look up the "Science and Technology" category.
4. Website: https://phet.colorado.edu/sims/html/circuit-construction-kit-dc/latest/circuit-construction-kit-dc_en.html)
   Webpage Title: Circuit Construction Kit: DC *(This website gives you opportunities to create circuits.)*

Q1e. What does "renewable energy" mean?

_____

_____

_____

_____

_____

_____

_____

_____

_____

_____

2. Imagine

Information

For lights to work, they need to be connected to a circuit. A circuit is a complete path through which electric current can flow.

A simple electric circuit consists of three components:

1. *Power source:* A power source such as a battery or solar panel provides electrical energy that delivers an electric current through the circuit.

2. *Conducting wires:* These are the pathways that carry the electric current from the power source to the components in the circuit. Wires are usually made from the metals copper or aluminium because they allow electric current to flow easily.

3. *Components:* These are the devices in the circuit that use and control the electric current. Examples include light bulbs, motors and switches.

Q1f. How does a solar panel work?

_____

_____

_____

_____

_____

_____

_____

Challenge 4. A solar solution for lighting spaces

 Simple circuits can be drawn using icons or symbols, for example:

Battery   Solar Panel   Wire   Bulb   Switch

Consider these simple circuits below and answer the following questions:

Q2a. When the switch is turned on, the bulb lights up. What is happening?

_____
_____
_____
_____

Q2b. When the switch is turned off, the bulb does not light up. Why?

_____
_____
_____
_____

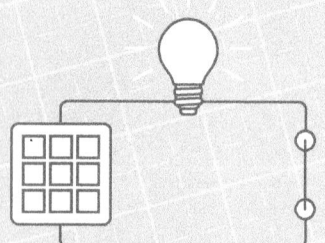

Q2c. In this circuit, the battery is replaced by a solar cell. When the switch is turned on, the bulb lights up. How does this circuit work?

_____
_____
_____
_____

Example a

Here are two examples of circuits. Each has two bulbs that are connected differently from solar cells.

Q2d. What is the difference between the two connections?
(Your teacher might also give you a chance to set up and test these circuits.)

_____

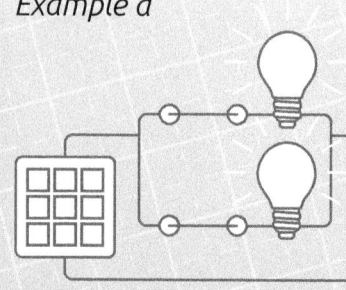

Example b

_____
_____
_____
_____

Challenge 4. A solar solution for lighting spaces

Q2e. How does a switch work?

Q2f. Copper or aluminium wires are usually covered with rubber or plastics. Why?

3. Plan

Q3a. Using the floor plan below draw a sketch of your circuit that you will set up to wire up your garden shed. Use icons or symbols to show all the components of your circuit.

Show your sketch to your classmates and get their feedback.

Q3b. What did your classmates think? Write their comments here:

Challenge 4. A solar solution for lighting spaces

Consider your classmates' feedback.

Q3c. Review your initial sketch of the circuit. Now, draw the final sketch of your circuit idea. Some things to address in the final design are:
- position of the lights
- position of the solar panel
- position of conducting wires and switches.

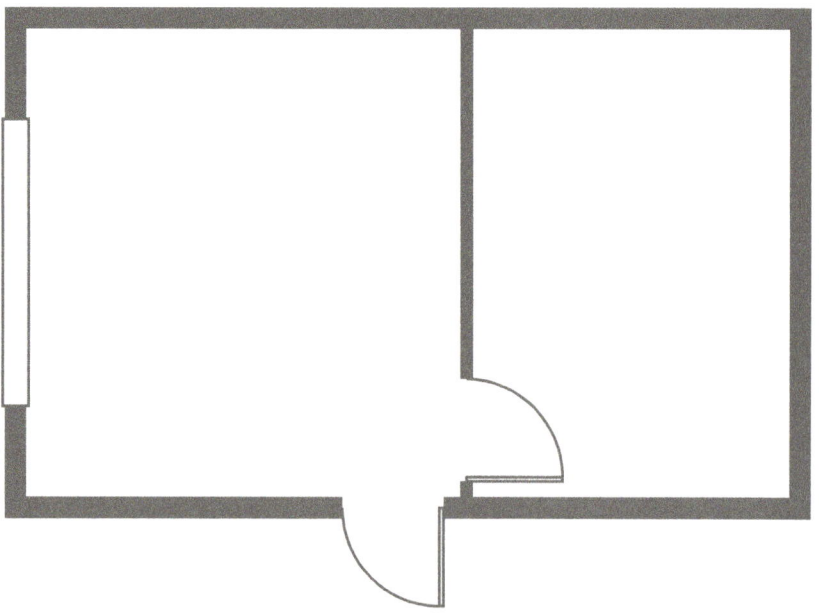

Q3d. What materials will you use to set up your circuit? Use the table below to list the materials, their purpose and the quantities you will need.

Material	Purpose	Quantity

Challenge 4. A solar solution for lighting spaces

Procedure

Q3e. What procedure will you follow to set up your circuit for the garden shed? Remember to include the steps for installing the solar panel, wires, switches and lights. And do not forget to include the steps for setting up the card box for the installation of the circuit.

Tools & Equipment

Safety is important to prevent accidents and injuries when working with electricity, and only qualified persons such as electricians using safety gear are allowed to work with mains power. The other term for safety gear is "Personal Protective Equipment" or "PPE". For every activity you undertake, you need to understand the safety rules and the possible consequences of not following them.

Q3f. List two safety rules that you need to follow when setting up a circuit. What are the consequences if they are not followed? List them in the table.

Safety rule	Consequences if the rule is not followed
1	
2	

Challenge 4. A solar solution for lighting spaces

4. Create

Creating and installing a circuit

Seek feedback from your teacher while you are conducting your experiment.

You need to:
- gather the components, materials and equipment
- follow your procedure
- follow the safety rules.

Q4a. Making your garden shed and installing the lights and solar panel may not go as planned. For each step in the procedure that caused problems, explain how you dealt with it. List the step(s) and your course of action.

Q4b. Test the lights in the garden shed. Are they working?

☐ Yes   ☐ No   ☐ Sort of/Unsure

"If No or Sort of/Unsure investigate the circuit and find the fault. How did you fix it?

Q4c. What are two **variables** to consider when installing solar panels?

1

2

Challenge 4. A solar solution for lighting spaces

5. Improve

Present your finished garden-shed model with lights to the class. Convince them that your idea aligns with the UN's goal and has addressed the challenge for this activity.

Ask your classmates these questions:
- *What are some of the good points about my finished model?*
- *What are some of the weak points about my finished model?*

Q5a. Write the most useful answers from your classmates in the table.

Good points

Weak points

Consider your classmates' answers

Q5b. What could you do to improve your finished model?

_____

_____

_____

Now it's time to reflect...

What have you learnt in this activity? *List three things.*
Hint: Look back at the key words.

1. _____

2. _____

3. _____

What would you like to know more about?

_____

_____

_____

_____

 *Try the quiz on Kahoot. See your teacher.*

 You can also use computer programs to show how circuits work. How can you design and program the circuit you proposed for the garden shed using a software program such as Scratch?

Need more space? Use pages 105–108.

Challenge 4. A solar solution for lighting spaces

The United Nations Sustainable Development Goal 11

Sustainable Cities and Communities

is about making cities and other human settlements safe, resilient, and sustainable. In a world of conflicts and wars, people flee their homelands in the hope they can live in other, more peaceful countries. These people are known as refugees. They need places to live when they arrive in a new country. One of the targets of Goal 11 is to ensure that everyone has access to affordable housing.

Key words in this activity:
- assembly tool
- design
- floor plan
- kit home
- refugee camp

A kit shelter for refugees

A building designer has challenged your class to **design** and create a scale model of a kit shelter for emergency accommodation. There is a need for cheap and appropriate shelters for the homeless and refugees fleeing areas of conflict in our world. Design a one-level kit shelter that considers the climate and the four seasons. Its size should be 5 metres × 3 metres and it is for a family of four: two adults and two children. The shelter can be made from modern materials and/or sustainable materials. The scale model will be displayed during Refugee Week to seek feedback from the public on the sustainability of your idea.

What is a building designer?

A building designer is someone whose job it is to design new buildings and prepare drawings for construction.

A scale model is a smaller, or larger, version of an item.

*There are many **refugee camps** around the world.*

The world's largest refugee camp is Cox's Bazar in Bangladesh.

1. Ask

Q1a. What is a kit shelter?

_____
_____
_____
_____

Q1b. Why do we need a kit shelter?

_____
_____
_____
_____
_____

Q1c. What should a kit shelter be like? *Complete the table* →

A kit shelter should be:	Why?
• easy to erect	
• easy to dismantle	
• cheap to build	
• easy to transport	
• easy to store	

r — Do some research in your library and on the Internet to find answers to the questions that follow. This may also help you to think through some of the other questions in this activity.

Here are some URLs to get your research underway:

1. Website: https://hipages.com.au/article/what_is_a_kit_home
   Webpage title: What is a kit home
2. Website: https://seelibrary.org/ Webpage Title: SEE library
   Look up the "Science and Technology" category.

Refugee camp in Cox's Bazar, Bangladesh

Q1d. Look at the photo above. Record two significant points about this refugee camp that catch your eye.

1 _____ *2* _____

Q1e. List three interesting things about **kit homes**.

1 _____ *2* _____ *3* _____

Q1f. What materials – natural or man-made – could be used to make a kit shelter? Why would you use them? Complete the table below.

Material	Why?

Challenge 5. A kit shelter for refugees

2. Imagine

Kit shelter design

Some things to consider when designing a kit shelter were listed in Q1b. Consider what type of structure *(see page 94, "What are structures")* would meet these requirements.

Remember, the size of the shelter is 5 metres × 3 metres. Space will be limited so every part of the shelter should have a purpose. Also, every surface should have a use, whether horizontal (e.g. ceiling) or vertical (e.g. wall).

An open-plan layout could be a choice. Flexible living areas are an option – for example, beds and tables could fold away or be raised when not in use. Indoors could be part of the outdoors. A side, or part of a side, could be removed or opened.

Shelters can be fixed or temporary structures. They could collect rainwater and store it. External walls could be areas to grow vegetable gardens, to dry clothing or crops, or to attach sustainable alternative energy source(s).

The orientation of a kit shelter to the Sun's path is important. There is a need to block light and heat during summer. An extended roof or an awning above a window or opening provides shade. However, during winter, light and warmth are required. Be aware also of shadows cast by other objects, especially on any surface used to grow plants, drying crops, and clothing.

The climate affects the way people live and what they can do around the world.

Earth is impacted by our planet's movement. Its daily rotation and yearly revolution around the Sun give us day, night, and seasons.

Seasons are the result of the Earth's tilt as it orbits the Sun.

23.5° tilt
Axis of Rotation

The Earth revolves Eastwards on its axis of rotation, with a tilt of 23.5 degrees.

Challenge 5. A kit shelter for refugees

Q2a. Sketch the Earth's position relative to the Sun for each season of your hemisphere.

Summer

Autumn

Winter

Spring

Shadows are formed by a solid object blocking light. The area of darkness is behind the object, that is opposite the light source.

Q2b. The size of the shadow varies as the seasons change. Explain why this is so in a sketch. *See page 56.*

Challenge 5. A kit shelter for refugees

3. Plan

Q3a. Use the grid below to draw a rough sketch of your initial idea in 2D or 3D for a kit shelter.

Show your sketch to your classmates and get their feedback.

Q3b. What did your classmates think? Write their comments here:

_____
_____
_____
_____
_____
_____
_____
_____

Consider your classmates' feedback.

Q3c. Review your initial sketch. Now, draw the final sketch of your kit-shelter idea. Remember to:
- label the materials, features and services
- add any other information and details.

Challenge 5. A kit shelter for refugees

i *Sunlight heats the ground and things on the Earth. As the position of the sun changes, the temperatures of the seasons also vary, and so do the size of the shadows. This diagram shows the shadows cast Summer and Winter in the Southern Hemisphere.*

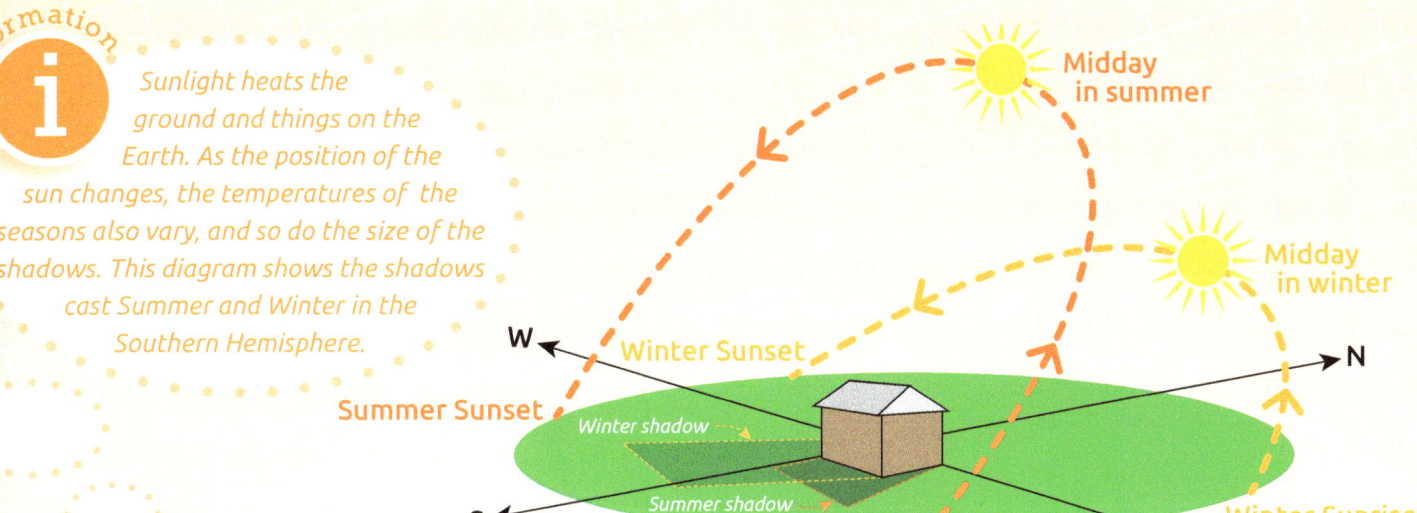

*Engineers/builders need much information to build structures. There is a **floor plan**. There are location and orientation drawing(s). There may also be other drawings, for example, landscaping and services. There is also a materials list – see Q3e next page.*

Q3d. Present a floor plan. You could use a design application (*see Curly Question, page 62*). Include a sketch to show how one should place the kit shelter. Indicate North, the Sun's pathway and shadows cast Summer and Winter. Show also the net of your 3D shaped kit shelter.

Q3e. In the table below, list the materials, their purpose, and the quantities you will need to make your kit-shelter model.

Material	Purpose	Quantity

Procedure

Q3f. Set up your procedure for constructing a kit shelter by drawing a line from each step number to the correct task.

Step	Task
1	Add the roof
2	Connect services/amentities
3	Erect the walls
4	Fit the doors and windows
5	Install the floor
6	Mark out the layout
7	Prepare the site
8	Select a site
9	Unpack the kit-shelter

Safety is important to prevent accidents and injuries. Using tools and equipment can be dangerous when assembling and erecting kit shelters. You should always wear safety gear for your protection. The other term for safety gear is "Personal Protective Equipment" or "PPE".

A variety of tools, known as assembly tools, can be used to erect a kit shelter.

*Some **assembly tools** are: a. Cordless drill; b Phillips head screwdriver; c. Spanner; d. Hammer; e. Spring clamp.*

Q3g. Select three tools from those shown above and write them in the "Assembly tool" column of the table below.

Q3h. Now complete the other two columns in the table:
• Every tool can injure you. Write down how they can injure you.
• What safety rules should you follow?

Assembly tool	How I could get injured	Safety rules I should follow

58

Challenge 5. A kit shelter for refugees

4. Create

Making your kit-shelter model

Seek feedback from your teacher while you are making your model.

You need to:
- gather materials
- gather the tools, equipment and adhesives you need
- follow your procedure to build your kit-shelter model
- use PPE and follow the safety rules.

Q4a. Making your kit-shelter model may not go as planned. For each step in the procedure that caused problems, explain how you dealt with it. List the step(s) and your course of action.

_____
_____
_____
_____
_____
_____
_____
_____
_____
_____
_____
_____
_____
_____
_____
_____

Q4b. It is time to review your design and test your solution. How well has your solution addressed the challenge? Record your comments here.

Q4c. How does the changing position of the Sun affect your model? *Hint: Shine a torch from different positions in a darkened area to observe and understand the effect of the Sun. Can you make any changes to your model in order to make the best use of the Sun?*

_____
_____
_____
_____
_____
_____
_____
_____
_____
_____
_____
_____
_____
_____
_____
_____
_____
_____

5. Improve

Present your kit-shelter idea to the class. Convince them that your idea aligns with the UN's goal and has addressed the challenge for this activity.

Ask your classmates these questions:
- What are some of the good points about my kit-shelter idea?
- What are some of the weak points about my kit-shelter idea?

Q5a. Write the most useful answers from your classmates in the table.

Good points

Weak points

Consider your classmates' answers

Q5b. What could you do to improve your kit-shelter idea?

_____
_____
_____

Now it's time to reflect...

What have you learnt in this activity? *List three things.*
Hint: Look back at the key words.

1. _____
2. _____
3. _____

What would you like to know more about?

_____
_____
_____
_____

 *Try the quiz on Kahoot. See your teacher.*

 Using a design application such as Key Plan 3D, create a floor plan and 3D model of your kit home. From these representations, consider what changes you could make to improve your kit-home model?

Need more space? Use pages 105–108.

The United Nations Sustainable Development Goal 13

Climate Action

*is about taking urgent action to combat climate change and its impacts. Changes in our climate can have adverse effects on everything that lives on our planet. What we do on Earth affects the climate. The dumping of **waste** causes pollution, which can affect the world's climate. We need to find more ways to **upcycle** wastes as this is an effective way to reuse waste for meaningful purposes. One of the targets of Goal 13 is to take action to tackle climate change.*

Key words in this activity:
- mechanisms
- motion
- upcycle
- waste

A toy that moves made from waste

Challenge 6

A toy designer has challenged your class to design and make toys from waste. Your toy, or part of it, must be able to move. The toy is for children at the local kindergarten. Your finished toy will be displayed during a recycling event such as Global Recycling Day or Recycle Week.

Information

A toy designer is someone whose job it is to design and create toys. A model of a toy is made and tested.

My plastic, my responsibility.
Your plastic, your problem.

My rubbish, my responsibility.
Your rubbish, your problem.

Children have always played with toys.
Toys have changed over time.

1. Ask

Q1a. Use the Internet to identify when each of these toys was made and write the date in the space provided. Mark the date on the timeline with a dot (●) and write the toy next to it. *An example is shown for LEGO.*

- Remote-controlled robot
- Transformers
- Tamagotchi
- Barbie doll
- Game Boy
- Yoyo

- LEGO 1949
- Monopoly
- Teddy Bear
- Rubik's Cube
- Hot Wheels cars
- Nintendo Entertainment System

1949 – LEGO

64

Challenge 6. A toy that moves made from waste

Q1b. How have toys changed over the last 100 years? _____

_____

_____

_____

_____

_____

_____

1

2

3

Motion is another word used to describe something that is moving. There are a number of ways toys can move. For example, roll, spin, rock, rotate, or a part can move.

Q1c. The photos above show a few toys that move. How do these toys move?

1 _____

2 _____

3 _____

Research

r   Do some research in your library and on the Internet to find answers to the questions that follow. This may also help you to think through some of the other questions in this activity.

Here are some URLs to get your research underway:

1. Website: https://www.youtube.com/watch?v=J0yntx0O07I
   Webpage Title: Moving toys

2. Website: https://www.stem.org.uk/elibrary/list/20488/gears-an-pulleys
   Webpage Title : Levers, Pulleys & Gears

3. Website: https://seelibrary.org/   Webpage Title: SEE library
   Look up Science and Technology category.

*Motion is achieved by using **mechanisms**. Mechanisms transmit force and movement to produce a result.*

Q1d. A few mechanisms are shown below. Use the Word Bank to identify each mechanism.

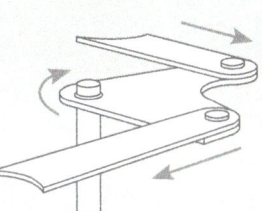 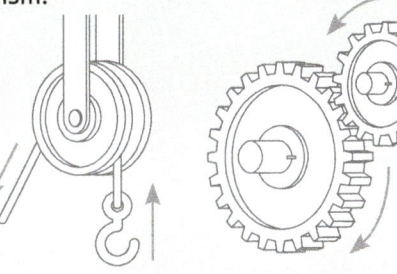 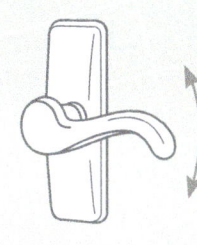

WORD BANK

gear   lever
linkage   pulley

_____ _____ _____ _____

Q1e. What other mechanisms are there? Use the Internet and list three.

1 _____

2 _____

3 _____

2. Imagine

Q2a. What type(s) of motion do you want your toy to have? Record your thoughts here.

_____
_____
_____
_____

Q2b. What mechanisms could you use to create the motion?

_____
_____

Q2c. What recyclable materials could you use to build your toy?

_____
_____

3. Plan

Q3a. Use the grid to draw a rough sketch of your initial idea for a toy that moves. Label the parts of your toy.

Show your initial toy idea to your classmates and get their feedback.

Q3b. What did your classmates think? Write their comments here:

_____
_____
_____
_____
_____
_____
_____
_____
_____
_____
_____
_____
_____

Consider your classmates' feedback.

Q3c. Review your initial sketch. Now, draw a final sketch of your toy that moves. Remember to:
- label the parts, materials and mechanism(s)
- show the movements and dimensions
- add any other information and details.

Q3d. In the table below, list the part and its purpose, the material you will use, and the quantities you will need to make your toy model.

Part and its purpose	Material	Quantity

Q3e. What procedure will you follow to make your moveable toy?

_____
_____
_____
_____
_____
_____
_____
_____
_____
_____

Tools & Equipment

Safety in toys is important to prevent accidents and injuries. Using tools and equipment can be dangerous. Be aware of things that could injure or harm children as they play with the toy.

Q3f. What are some of the safety issues with children's toys? List them below.

_____
_____
_____
_____
_____
_____

Compare your safety issues with some classmates. Add any other safety issues to your list.

A range of hand tools and power tools can be used to build toys.

Some of the tools could be: a. Cordless drill; b. Industrial snips; c. Warrington hammer; d. General-purpose saw; e. Phillips head screwdriver.

Q3g. Select three tools from those shown above and write them in the "Tool" column of the table below.

Q3h. Now complete the other two columns in the table:
• Every tool can injure you. Write down how they can injure you.
• What safety rules should you follow?

Tool	How I could get injured	Safety rules I should follow

4. Create

Making a toy that can move

Seek feedback from your teacher while you are creating your toy model.

You need to:
- gather the materials and equipment
- gather the tools you need
- follow your procedure for making the toy
- use PPE and follow the safety rules.

Q4a. Making your toy that moves may not go as planned. For each step in the procedure that caused problems, explain how you dealt with it. List the step(s) and your course of action.

_____
_____
_____
_____
_____
_____
_____
_____
_____
_____
_____
_____
_____

Challenge 6. A toy that moves made from waste

Test your design

Q4b. How well does your toy work? Record your comments here.

Do a safety-in-toy review

Q4c. Complete the questions below. Identify any safety concerns and suggest ways to rectify them.

- Are there any small parts? ☐ No ☐ Yes

  What are the safety concerns? _____   I will rectify this by: _____

  _____   _____

  _____   _____

- Are there any projectiles? ☐ No ☐ Yes

  What are the safety concerns? _____   I will rectify this by: _____

  _____   _____

  _____   _____

- Are moving parts uncovered? ☐ No ☐ Yes

  What are the safety concerns? _____   I will rectify this by: _____

  _____   _____

  _____   _____

- Are there sharp edges and/or points? ☐ No ☐ Yes

  What are the safety concerns? _____   I will rectify this by: _____

  _____   _____

  _____   _____

- Are the materials toxic or flammable? ☐ No ☐ Yes

  What are the safety concerns? _____   I will rectify this by: _____

  _____   _____

  _____   _____

- Are there parts that can break easily? ☐ No ☐ Yes

  What are the safety concerns? _____   I will rectify this by: _____

  _____   _____

  _____   _____

5. Improve

Present your toy idea to the class. Convince them that your idea aligns with the UN's goal and has addressed the challenge for this activity.

Ask your classmates these questions:
- *What are some of the good points about my moveable toy?*
- *What are some of the weak points about my moveable toy?*

Q5a. Write the most useful answers from your classmates in the table.

Good points

Weak points

Consider your classmates' answers

Q5b. What could you do to improve your moveable-toy idea?

_____
_____
_____
_____
_____

Now it's time to reflect...

What have you learnt in this activity? *List three things.*
Hint: Look back at the key words.

1. _____

2. _____

3. _____

What would you like to know more about?

_____
_____
_____
_____

Try the quiz on Kahoot. See your teacher.

Curly Question

Prepare a poster that draws attention to our consumerism and energy consumption. Choose one of these options:

Option A: Use art materials and equipment.

Option B: Use digital technologies and an application such as Canva.

My choice is: ◯ Option A   ◯ Option B

My STEM Workbook 3 – Understanding Science, Technology, Engineering and Mathematics through design-process activities   Vinesh Chandra and Basil Slynko   ISBN: 978-0-6484052-2-1

The United Nations Sustainable Development Goal 14

Life below Water

is about conserving and sustainably using the oceans, seas and marine resources. Products made of plastic represent most of the wastes that end up in our waterways. When these products are not disposed of properly, there is a chance that they can end up in lakes, streams, rivers and oceans. Plastics in waterways impact marine life. One of the targets of Goal 14 is to reduce marine **pollution** by removing wastes such as plastics for a healthier marine life.

Key words in this activity:

- algorithm
- codes
- computational thinking
- computer game
- plastics
- pollution
- program
- robot
- Sprite

Challenge 7. A video game on plastic-waste removal

A video game on plastic-waste removal

A video-game designer has challenged your class to design and **program** a computer game in which a **robot** removes plastics from a marine environment. The robot in your game should be able to move around and grab rubbish. Each time this happens the robot is rewarded with a point. Your teacher would like you to create this computer game using Scratch software. The game also needs a score counter. Your teacher would also like you to share your computer game on the Scratch community.

1. Ask

Q1a. Why do we need clean waterways?

_____

_____

_____

_____

_____

_____

_____

Video game designers design, create and produce video games.

Q1b. How do computer games work?

_____
_____
_____
_____
_____
_____

Q1c. The application of **computational thinking** is very important in coding or computer programming **computer games**. What does this mean?

_____
_____
_____
_____
_____
_____

Q1d. Research shows that in the past 30 years, there has been a sharp increase in the use of **plastics**. What is the reason for this increase?

_____
_____
_____
_____
_____

Q1e. Why should rubbish be removed from waterways?

_____
_____
_____
_____
_____

Q1f. What is a robot?

_____
_____
_____
_____
_____

Research

Do some research in your library and on the Internet to find answers to the questions that follow. This may also help you to think through some of the other questions in this activity.

Here are some URLs to get your research underway:

1. Website: https://www.wwf.org.au/news/blogs/the-lifecycle-of-plastics#gs.ojrjtf
   Webpage Title: The lifecycle of plastics
2. Website: https://www.kidsnews.com.au/environment/kids-tackle-water-pollution-with-found-object-prototypes/news-story/8e5073bd1b351e360b9e9656816038a4
   Webpage Title: Kids tackle water pollution
3. Website: https://seelibrary.org/ Webpage Title: SEE library
   Look up the "Science and Technology" category.

Challenge 7. A video game on plastic-waste removal

Q1g. What information did you gather about rubbish such as plastics and water pollution? List the three main points.

1 _____ 2 _____ 3 _____

2. Imagine

Q2a. How can we remove rubbish from our waterways? Suggest three ideas.

1 _____

2 _____

3 _____

Q2b. How can a robot be used to remove rubbish from waterways?

_____

Q2c. Why should a robot be used instead of a person?

_____

Watch a video on making a computer game in Scratch: see the workbook website http://mystemworkbook.com

The Scratch home page screen of a computer program created by Marta is shown above. The robot is controlled by control keys (up, down, left, right) on the keyboard. Rubbish randomly falls from the top of the screen. The game player needs to move the robot so that it can 'grab' the rubbish. Each time the robot collides with an item of rubbish, the player gets a point. Link to Scratch website https://scratch.mit.edu/

Q2d. What are some of the strengths and weaknesses of Marta's game?

Strengths	Weaknesses

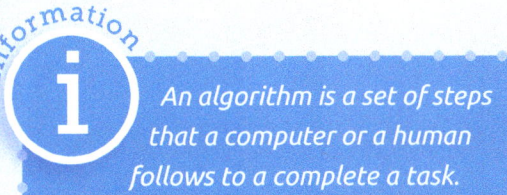

An algorithm is a set of steps that a computer or a human follows to a complete a task.

Visit the Scratch website. Search and play some games.

Q2e. Jot down some points that will help you create your game. *Hint: Search "catch rubbish" on the Scratch website.* If you sign in to the site, you will be able to see the algorithms that were used to create the game.

_____
_____
_____
_____
_____
_____
_____
_____
_____
_____
_____
_____
_____
_____

3. Plan

Q3a. Draw a rough sketch of the home page of your "Scratch" game.

Show your sketch to your classmates and get their feedback.

Q3b. What did your classmates think? Write their comments here:

_____
_____
_____
_____
_____
_____
_____
_____
_____

Consider your classmates' feedback.

Q3c. Review your initial sketch. Now, draw the final sketch of the home page of your "Scratch" game. Label the following in your sketch:
- robot **sprite**
- other sprites
- score counter.

Q3d. In the table, list the materials (software and hardware) and their purposes in making the game.

Hardware and software requirements	Purpose

Procedure

Q3f. What procedure will you follow to make your game? Write the steps here.

Tools & Equipment

Safety is important when using computers and the Internet.

Q3e. What are the three main safety rules and what are the consequences if they are not followed? List them below.

Safety rule 1
Consequence if the rule is not followed

Safety rule 2
Consequence if the rule is not followed

Safety rule 3
Consequence if the rule is not followed

4. Create

Making a computer game

Seek feedback from your teacher while you are creating your computer game.

You need to:
- log in to your computer
- open the "Scratch" interface
- follow the rest of your procedure
- follow the Internet safety rules.

Q4a. Making your computer game may not go as planned. For each step in the procedure that caused problems, explain how you dealt with it. List the step(s) and your course of action.

_____
_____
_____
_____
_____
_____
_____
_____
_____
_____
_____
_____
_____

Q4b. Conduct a survey with five participants and gather feedback on your game. The participants could be your mates, siblings, parents and grandparents.

Ask them the following questions and record their answers in the table using tally marks (|), then write the total in the shaded column:

a. How was the design of the robot sprite in my game?
b. How was the backdrop in my game?
c. How was the algorithm in my game?
d. How was my game overall?

	Good	Average	Needs work
	Total	Total	Total
Question a			
Question b			
Question c			
Question d			

Once you have conducted the survey, analyse the results. Use the totals from the table above and show the results in the graph below. Remember to add labels.

Label (Y axis): _____

Label (X axis): _____

Use the survey results to improve your game before you present it to the class.

5. Improve

Present your computer game to the class. Convince them that your idea aligns with the UN's goal and has addressed the challenge for this activity. Explain the **algorithm** in your game.

Ask your classmates these questions:
- What are some of the good points about my video game?
- What are some of the weak points about my video game?

Q5a. Write the three most useful answers from your classmates in the table.

Good points	Weak points

Challenge 7. A video game on plastic-waste removal

Consider your classmates' answers

Q5b. What could you do to improve your video game before you upload it to the "Scratch" website?

_____

_____

_____

Now it's time to reflect...

What have you learnt in this activity? *List three things.*
Hint: Look back at the key words.

1. _____

2. _____

3. _____

What would you like to know more about?

_____

_____

_____

_____

 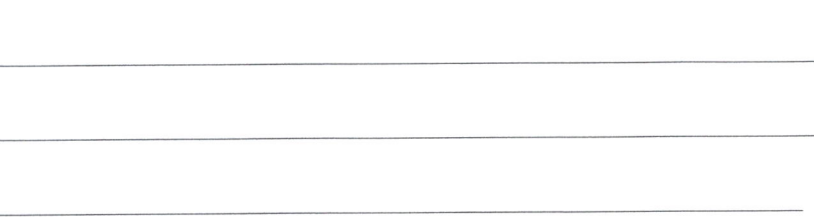

Try the quiz on Kahoot. See your teacher.

 What **codes** can you modify in your "Scratch" program so that the robot is automated and picks up rubbish by itself without the need for human control?

Need more space? Use pages 105–108.

My STEM Workbook 3 – Understanding Science, Technology, Engineering and Mathematics through design-process activities   Vinesh Chandra and Basil Slynko   ISBN: 978-0-6484052-2-1

The United Nations Sustainable Development Goal 15

Life on Land

is about protecting, restoring, managing and promoting the sustainable use of **ecosystems**. Caring for the life of all living things on Earth is important. They are a part of our ecosystem. All living things need shelter. One of the targets of Goal 15 is to take action to protect wildlife. Shelters protect living things from the weather and **predators**. They are also a place to raise their young.

Key words in this activity:

- design • ecosystem
- floor plan • form
- frame • habitat
- predators • shell
- specifications
- structures • vertices

Challenge 8. A habitat for giraffes

A habitat for giraffes

Challenge 8

A habitat specialist has challenged your class to **design** and make a model of the new habitat. The habitat, including a shelter, is for a family of giraffes: a bull, a cow, and a calf. Suitable models will be displayed in the zoo's foyer to seek feedback from the public on the suitability of your design.

What does a habitat specialist do?

A habitat specialist is someone whose job it is to design and build homes for animals in zoos and protected areas.

1. Ask

Q1a. Why is caring for wildlife important?

_____
_____
_____
_____
_____

Q1b. What does the term 'habitat' mean to you?

_____
_____
_____
_____
_____

> **Information**
> When designing a habitat you need to consider factors such as the wellbeing of the zookeepers and the public, and the welfare of the giraffes.

Q1c. With these factors in mind, what are some of the specific requirements of designing a habitat for a zoo? Record your thoughts.

Wellbeing of the zookeepers:

_____
_____
_____
_____
_____
_____

Wellbeing of the public:

_____
_____
_____
_____
_____
_____

Welfare of the giraffes:

_____
_____
_____
_____
_____
_____
_____

> **Research**
> Do some research in your library and on the Internet to find answers to the questions that follow. This may also help you to think through some of the other questions in this activity.

Here are some URLs to get your research underway:

1. Website: https://www.natgeokids.com/au/discover/animals/general-animals/ten-giraffe-facts   Webpage title: Giraffe shelters for kids
2. Website: https://www.dezeen.com/2014/06/10/monk-mackenzie-and-glamuzina-patterson-create-angular-giraffe-shelter-at-auckland-zoo/amp
Webpage title: Dezeen giraffe shelters
3. Website: https://seelibrary.org/   Webpage Title: SEE library
Look up the "Science and Technology" category.

Q1d. Your habitat design could be geometric or organic in **form**. Explain these terms.

Geometric form is: _____

_____

_____

_____

Organic form is: _____

_____

_____

_____

Q1e. What are the dimensions of a giraffe? What is the average height of an adult male and female giraffe as well as the average body length from chest to tail?

Average height of a male giraffe is:

_____

Average height of a female giraffe is:

_____

Average body length (chest to tail) is:

_____

Q1f. Why is knowing information about height and length important?

_____

_____

_____

_____

_____

_____

_____

_____

_____

_____

_____

_____

Height

Body Length

Q1g. What features are needed for a giraffe habitat?

_____

_____

_____

_____

_____

_____

2. Imagine

The welfare of the giraffes is one of the design specifications. The giraffes therefore will require a shelter within the habitat.

Q2a. What should the minimum size of the shelter be within the habitat for the family of giraffes? Insert the dimensions on the diagram below.

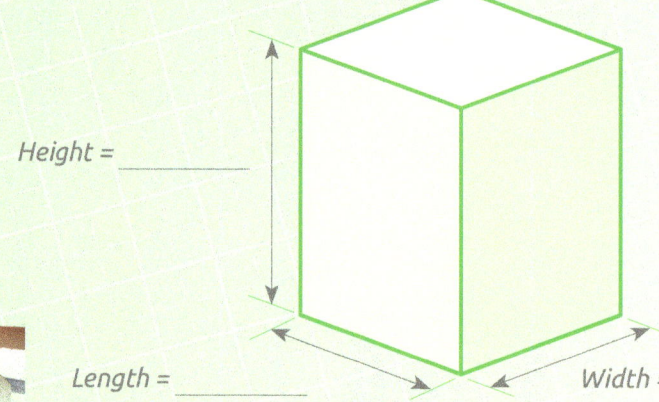

Height = _____

Length = _____   Width = _____

Q2b. What size should the habitat be for the family of giraffes? Sketch a few shapes that the habitat could be and add some basic measurements.

The shapes of the habitat could be:

_____

_____

_____

Now compare your dimensions with those of other students.

Q2c. Do you need to adjust the size of the shelter and/or habitat?

◯ Yes   ◯ No   ◯ Maybe

Why/Why not? _____

_____

_____

_____

_____

Q2d. What roof design could you use on your shelter, an organic or a geometric form? Explain your answer.

_____

_____

_____

_____

_____

_____

Organic shapes, unlike geometric shapes, are free-flowing shapes as seen in nature – often irregular and asymmetrical.

The roof forms of the Sydney Opera House is one example of organic shapes used in architecture.

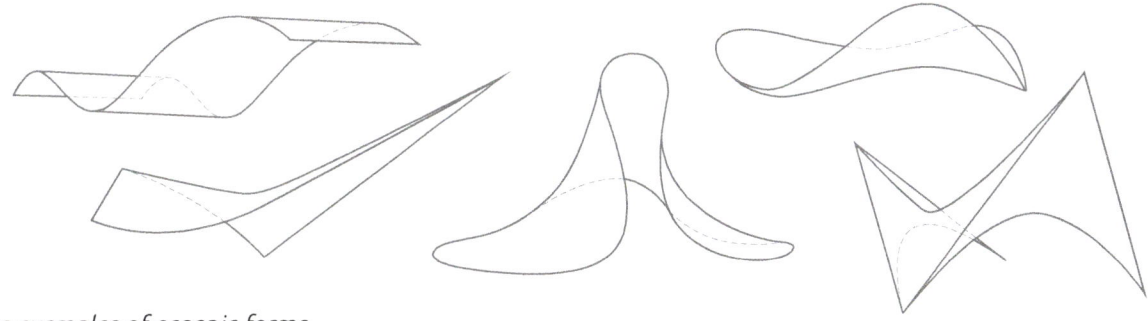

Some examples of organic forms.

Challenge 8. A habitat for giraffes

Q2e. Sketch a few organic roof designs.

Q2f. Several geometric roof designs are illustrated here. Use the Word Bank to identify the 3D shapes.

WORD BANK
Cone
Hemisphere
Rectangular prism
Square-based pyramid
Triangular prism

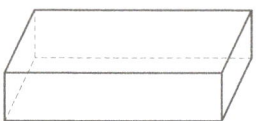

1. _____

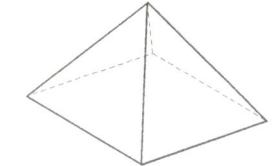

2. _____

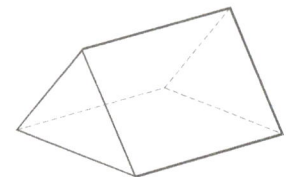

3. _____

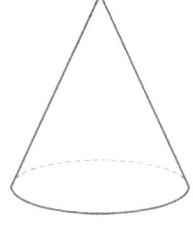

4. _____

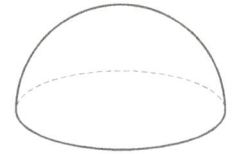

5. _____

Did you know that materials can be sorted into two basic groups. They are natural and man-made; that is, processed, synthetic and composite.

For more information, see page 104, "Some samples of natural and man-made materials".

The properties of a material, such as toughness, absorbency, hardness or brittleness, determine how the material should be used.

For more information, see page 103, "Some properties of materials".

Q2g. What materials could you use to build the shelter? List two suitable materials for walls, roof and floor in the "Suitable material" column of the table. Then write "W" for walls, "R" for roof and "F" for floor after each material. Next, identify the basic group to which each material belongs. *See "Some samples of natural and man-made materials", page 104.* Now add a Main property and a Weakness to each column.

Suitable material	Material group (natural/man-made)	Main property	Weakness
1			
2			
3			
4			
5			
6			

What are *structures*?

A shelter is a structure. A structure is an object that can carry a load and withstand other forces acting on it. There are two types of structures:
- a *frame*
- a *shell*.

Structures are built to:
- provide shelter, e.g. buildings
- control the environment, e.g. dams
- carry loads, e.g. bridges.

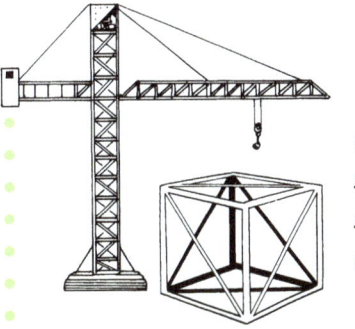

Frame structures: a frame structure made of triangular shapes. A triangle is a more rigid (not easily bent) shape.

Shell structures have no frame, but they have an outside skin.

3. Plan

Q3a. Use the grid to draw a rough floor plan, including basic measurements of your giraffe-habitat idea.

Then show your floor plan to your classmates and get their feedback.

Q3b. What did your classmates think of your **floor-plan** idea? Write their comments here.

_____
_____
_____
_____
_____
_____
_____
_____
_____
_____
_____

Consider your classmates' feedback.

Q3c. Review your rough floor plan. Now, draw the final floor plan of your giraffe-habitat idea. Remember to:
- label the materials, services and show the dimensions
- label the angles at the **vertices**
- sketch any specific details such as 3D view, information signs and warning signs
- add any other information and public safety details.

Challenge 8. A habitat for giraffes

Q3d. What materials will you use to make your giraffe habitat? Use the table below to list the materials, their purpose and the quantities you will need.

Material	Purpose	Quantity

Q3e. Calculate the total floor area of your giraffe habitat.

The total area is: _____

Q3f. Calculate the percentage of shelter area within the habitat.

The percentage of shelter area is: _____

Q3g. What procedure will you follow for making your giraffe habitat? Draw lines to identify the correct task for each step number.

Step	Task
1	Add fencing
2	Add roof
3	Apply finish
4	Build structure/shelter
5	Get materials
6	Install features
7	Mark out the habitat
8	Supply services

Challenge 1. A habitat for giraffes

Safety is important to prevent accidents and injuries. Using tools and equipment to build a giraffe habitat can be dangerous. You should always wear safety gear to protect yourself. The other term for safety gear is "Personal Protective Equipment" or "PPE". For every activity you undertake, you need to understand the safety rules and the possible consequences of not following them.

Some examples of PPE: a hard hat, protective eyewear and a hi-vis shirt.

Q3h. Identify your safety concerns if you were using the following tools at home to make a model.

Tool/equipment	Safety concern
Cordless drill	
Craft knife	
Cordless glue gun	
Super glue	
General-purpose saw	
Industrial snips	

Challenge 1. A habitat for giraffes

4. Create

Making your model of a giraffe habitat

Seek feedback from your teacher while you are making your model.

You need to:
- gather the materials including a baseboard
- gather the tools, equipment and adhesives
- follow your procedure to build your giraffe habitat model
- use PPE and follow the safety rules.

Q4a. Making your model of your giraffe habitat may not go as planned. For each step in the procedure that caused problems, explain how you dealt with it. List the step(s) and your course of action.

_____
_____
_____
_____
_____
_____
_____
_____
_____
_____
_____
_____
_____

Presenting your model of the giraffe habitat for display:

You need to:
- inspect your model for quality
- adjust if required
- apply a finish
- use paint pots to colour it
- do a final inspection
- prepare a baseboard.

Or use 3D modelling software. See Curly Question Option C, p101.

Evaluate your solution

Q4b. How well has your solution addressed the challenge?

_____
_____
_____
_____
_____
_____
_____
_____
_____
_____
_____
_____

5. Improve

Present your idea for the giraffe habitat to the class. Convince them that your idea aligns with the UN's goal and has addressed the challenge for this activity.

Ask your classmates these questions:
- *What are some of the good points about my giraffe-habitat idea and model?*
- *What are some of the weak points about my giraffe-habitat idea and model?*

Q5a. Write the most useful answers from your classmates in the table.

Good points

Weak points

Consider your classmates' answers

Q5b. What could you do to improve your giraffe habitat?

_____
_____
_____
_____

Now it's time to reflect...

What have you learnt in this activity? *List three things.*
Hint: Look back at the key words.

1. _____

2. _____

3. _____

What would you like to know more about?

_____
_____
_____
_____

q *Try the quiz on Kahoot. See your teacher.*

Curly Question

Choose one or more of the options below.

Option A: Is there a future for zoos?   ◯ Yes   ◯ No   ◯ Maybe
Record your thoughts on the next page.

Option B: Using a design application such as Key Plan 3D show the interior layout of your shelter. You can include this drawing as part of your giraffe habitat for display to the public.

Option C: Using free 3D modelling software, FreeCAD or Sketchup show the interior layout of your shelter to accompany your model.

Need more space? Use pages 105–108.

Some properties of materials

Different materials are used for different things. Have you ever wondered why? It is to do with the properties of materials.

Each material has its own set of properties. These properties determine how a material should be used. For example, wood is strong when squeezed (compression) and steel is strong when stretched (tension).

Hardness
How well does it resist denting or scratching?

Flammability
How easily does it burn or smoulder when exposed to heat or fire?

Strength
How easy or difficult is it to break?

Density
How heavy or light is it for its size?

Elasticity
How well does it return to its original shape and size after being subjected to external loads and forces?

Toughness
How much force can it stand without breaking?

Conductivity
How well does it conduct or insulate against heat, electricity or sound?

Durability
How well does it resist environmental stresses such as weather or insects, or does it tear easily or wear out (e.g. fabric)?

Notes

Notes

Notes

Notes

www.ingramcontent.com/pod-product-compliance
Ingram Content Group UK Ltd.
Pitfield, Milton Keynes, MK11 3LW, UK
UKHW061212180426
11947UKWH00028B/2006